STUDY GUIDES
Science
Year
5

Alan Jarvis and
William Merrick

RISING★STARS

Rising Stars UK Ltd, 22 Grafton Street, London W1S 4EX

www.risingstars-uk.com

Published 2007

Reprinted 2008

Reprinted 2009

Text, design and layout © Rising Stars UK Ltd.

Design: HL Studios and Clive Sutherland

Illustrations: Bookmatrix (Beehive Illustration)

Editorial project management: Dodi Beardshaw

Editorial: Marieke O'Connor

Cover design: Burville-Riley Partnership

British Library Cataloguing in Publication Data.

A CIP record for this book is available from the British Library.

ISBN: 978-1-84680-097-9

Printed by: Craft Print International Limited, Singapore

Contents

How to get the best out of this book

Each topic spreads across two pages and focuses on one major idea. Many of your lessons may be based on these topics. Each double page helps you keep **On track** and **Aiming higher**.

Title and key ideas: tell you what you are aiming to learn. The second idea is always more difficult than the first.

Key information: sets out the key facts that you need to know and the ideas you need to understand fully.

Key questions: The information in this section helps you learn more facts and understand the science in each topic. The investigations you do will give you the evidence you need to prove the scientific facts you've learnt.

Key words and their meanings: help build up your scientific vocabulary. Remember that some words mean one thing in everyday life and something more special in science.

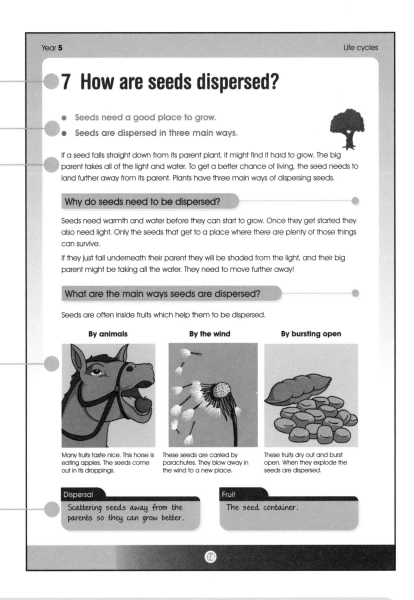

Follow these simple rules if you are using the book for revising.

1 Read each page carefully. Give yourself time to take in each idea.

2 Learn the key facts and ideas. Ask your teacher or mum, dad or the adult who looks after you if you need help.

3 Concentrate on the things you find more difficult.

4 Only work for about 20 minutes or so at a time. Take a break and then do some more work.

The right-hand page has lots of fun questions for you to try. They are like the National Curriculum test questions you will do. The answers are in the pull-out section in the middle of this book.

If you get most of the **On track** questions right, then you know you are working at level 4. Well done – that's brilliant!

If you get most of the **Aiming higher** questions right, you are working at the higher level 5. You're really doing well!

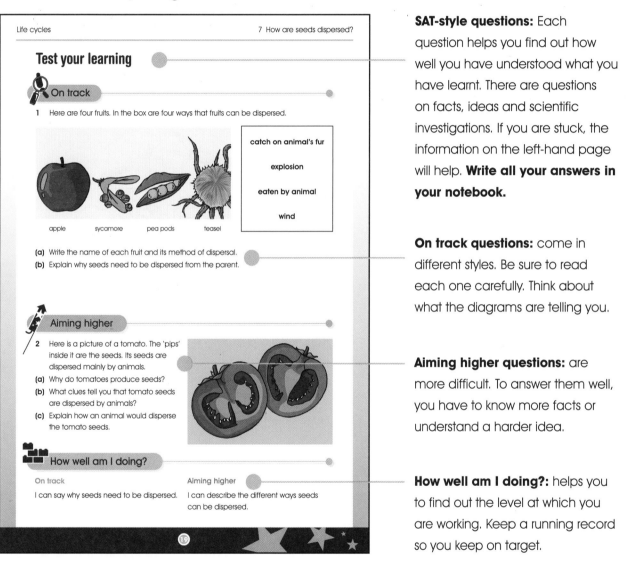

SAT-style questions: Each question helps you find out how well you have understood what you have learnt. There are questions on facts, ideas and scientific investigations. If you are stuck, the information on the left-hand page will help. **Write all your answers in your notebook.**

On track questions: come in different styles. Be sure to read each one carefully. Think about what the diagrams are telling you.

Aiming higher questions: are more difficult. To answer them well, you have to know more facts or understand a harder idea.

How well am I doing?: helps you to find out the level at which you are working. Keep a running record so you keep on target.

Follow these simple rules if you want to know how well you are doing.

1 Work through the questions.

2 Keep a record of how well you do.

3 If you are working at level 4 you will get most of the **On track** questions correct.

4 If you are working at level 5 you will also get most of the **Aiming higher** questions correct.

1 What is healthy eating?

- To stay healthy you need to eat the right food.
- Some evidence for this was found over 250 years ago.

Different foods do different things in your body. Some help you grow, some give you energy and some keep illness away. You must eat the right amount of each food – too much is as bad as too little. We need to eat many different foods each day to keep healthy.

What makes a healthy diet?

Meat, fish, eggs and cheese are good for building and repairing your body.

Sugar, carbohydrates like potatoes, fats and oils give you energy to keep warm and to move.

Fruits and vegetables help to keep you well.

You need some from each section to keep you healthy.

How was this discovered?

Explorers and sailors were the first people to notice that they got ill if they did not get enough fruit.

Sailors on long journeys used to get an illness called **scurvy**. They felt weak and tired. Their teeth fell out.

At sea they just lived on salted meat and dry biscuits for weeks on end. When they ate some fresh fruit ashore, they got better.

In 1753 Dr James Lind wanted to help the sailors. He gave them oranges on the voyage and they did not get scurvy. Lemons and limes worked too. This is because fruit contains a chemical called vitamin C, which helps to keep you healthy.

Diet

What we eat and drink.

Balanced diet

Diet which contains healthy amounts of all different foods.

Test your learning

On track

1   Here are four foods that Dominic likes.

pasta   chicken   apple   crisps

(a) Which one helps him grow bigger muscles?

(b) Which one will give him the most energy for running?

(c) Which one will help to stop him getting ill?

Aiming higher

2   Dr Lind had an idea that oranges and limes would help to cure scurvy, so he tried
     it out on some ill sailors. At the same time he tested other things. Some sailors were
     given cider, some lemons, some sea water and others got vinegar.

(a) Why would oranges and limes prevent scurvy?

(b) Explain why he included the other things in his test.

How well am I doing?

On track

I can say what each part of a healthy diet
is needed for.

Aiming higher

I can explain how Dr Lind discovered how
to cure scurvy.

2 What does your heart do?

- Your heart is in your chest; it is protected by your ribs.
- Your heart pumps blood around every part of your body.

Inside your chest, protected by your ribs, is your heart. It pumps blood to every part of your body in tubes called **blood vessels**. The blood goes all around the body and back to the heart, carrying energy to every part.

Where is the heart?

Jaldev's heart, like yours, is inside his chest cavity. It is about the size of his fist. The ribs help protect it from damage.

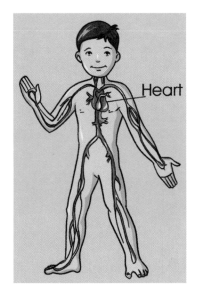

Heart

What does the heart do?

Jaldev's heart has spaces inside to hold blood. The muscular walls of his heart squeeze the blood and pump it around his body.

Blood circulates round and round his body, calling in on all of his major organs. It travels in tubes called blood vessels.

You might be able to see your blood vessels through the skin on the back of your hand or inside your wrist.

Blood vessel
A tube that carries blood.

Muscle
A body tissue that can pull or squeeze.

Test your learning

On track

1 Annetta and Jolie have got a model of the human heart.

(a) What is a real heart made of? (Pick **one** of these.)

| muscle | hair | blood | bone |

(b) What does the heart do to the blood?

(c) What word means that the blood goes round and round your body? (Pick **one** of these.)

| flows | circulates | runs | beats |

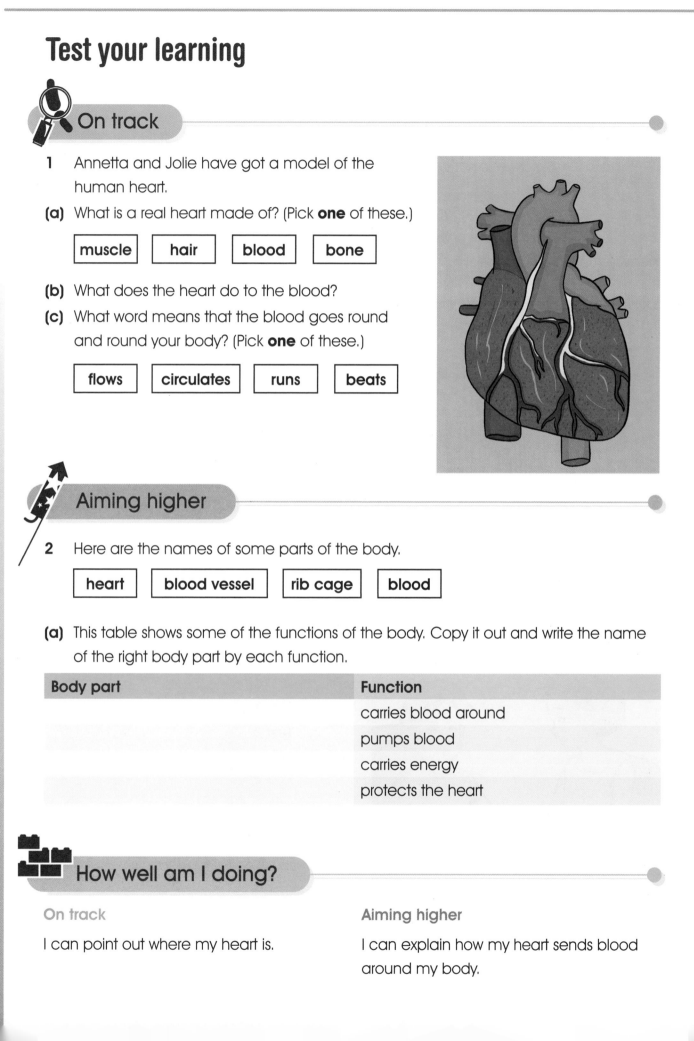

Aiming higher

2 Here are the names of some parts of the body.

| heart | blood vessel | rib cage | blood |

(a) This table shows some of the functions of the body. Copy it out and write the name of the right body part by each function.

Body part	Function
	carries blood around
	pumps blood
	carries energy
	protects the heart

How well am I doing?

On track

I can point out where my heart is.

Aiming higher

I can explain how my heart sends blood around my body.

3 Are exercise and pulse rate linked?

- Pulse rate is a measure of how fast your heart is beating.
- Single measurements are not always accurate.

Your pulse is caused by your heart beating. Your pulse rate speeds up when you take exercise. This delivers the extra energy your muscles need to work. To get a really accurate measurement of your pulse, you need to take several readings and work out an average.

What does the pulse rate tell us?

- The number of pulse beats in a minute is the number of times our heart beats.
- Your pulse rate might be 100 beats a minute; a grown-up's might be 70.
- Exercise makes your heart beat faster to send more blood to your muscles.

- Exercise makes you feel hot because the muscles are working hard.
- You also feel tired because you have used up some of your energy.
- After a rest you cool down and your pulse rate goes back to normal in a few minutes.

Why should you take an average?

Jolie sat down quietly and Dominic took her pulse. He took the reading five times.

	Pulse rate (beats/minute)
Test 1	100
Test 2	102
Test 3	98
Test 4	102
Test 5	98
Average	**100**

Each time it varied a little. The average result gives a better idea than each separate one.

Heartbeat

You can feel the heart pumping the blood if you feel your chest.

Pulse rate

The number of pulse beats in one minute is your pulse rate.

Test your learning

On track

1   Annetta wanted to find out
    what happens to her pulse
    rate when she runs.
    She wore a counter on
    her wrist to measure this.
    She started running, and
    stopped after ten minutes.

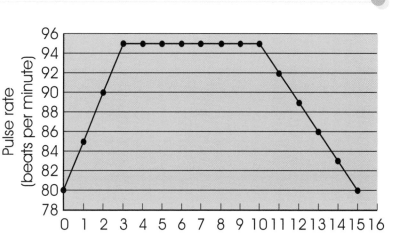

(a) What was Annetta's heart rate while she was running along?

(b) How long did it take her heart rate to go back to normal?

(c) Explain why her heart rate went up while she was running.

Aiming higher

2   Annetta repeated the test
    with five of her friends. She
    recorded each person's
    exercising pulse rate.

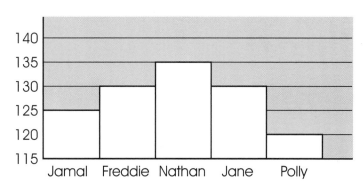

(a) What is the average exercising pulse rate of the five children in the bar chart?

| 136 | 122 | 128 | 130 |

(b) Explain why Annetta needed to take readings from five people.

How well am I doing?

On track

I can explain that your pulse rate shows
how fast your heart is beating.

Aiming higher

I can explain why measurements should
be taken from several people.

4 How do you keep healthy?

- To stay fit you need to take exercise and eat a healthy diet.
- You should also avoid taking poisons into your body.

Taking regular exercise will keep your heart and blood vessels in good condition. You must avoid harming yourself by taking in poisons. Tobacco smoke and alcohol are very dangerous. There are also many illegal drugs that are very poisonous.

How can you keep your heart and blood vessels healthy?

Jaldev made up some good rules for all the class to see.

- Eat plenty of fruit and vegetables. You need five portions a day.
- Don't eat so much food that you get overweight. This puts a strain on your heart.
- Avoid fatty foods. They cause a chemical to form in your blood that can stop it flowing.
- Take exercise to make your muscles and heart stronger and keep you at a healthy body weight.

What poisons should you avoid?

Tobacco

Smoking is very dangerous.

It will give you heart disease, and lung diseases such as bronchitis and lung cancer.

Over 100,000 people a year in the UK die young from smoking diseases.

Alcohol

Drinking alcohol stops a person thinking clearly and that may make them crash their car. It can also poison you.

Alcohol consumption kills about 50,000 people a year and this figure is increasing.

Other drugs

Illegal drugs such as heroin and cocaine are so dangerous you should never take them.

Most people avoid drugs, but 1000 people a year still die from taking them.

Heart attack

The heart stops beating because its blood vessels are blocked up.

Lung cancer

A disease caused by tobacco smoke.

Test your learning

On track

1   Here are some changes people can make to be more healthy.

| stop smoking | eat less fatty food | take more exercise | don't drink alcohol |

(a) Copy this table and match up each change with its benefit.

Change	Benefit
	Heart grows larger and stronger.
	Less chance of lung cancer.
	Become less fat.
	Less chance of a car crash.

(b) Explain how eating many fatty foods damages the heart.

(c) Dominic wants his dad to stop smoking. Write two reasons he can give him to explain why he should stop.

Aiming higher

2   A man is walking home one night after drinking too much alcohol. He is wobbling as he walks along, and it is hard to understand what he is talking about. He can't get his key into his door lock, and he is feeling a bit sick. Once inside, he soon falls asleep.

(a) Which of these is the most likely explanation for these things? Explain your answer.

| It is late at night and he is very tired. |
| The alcohol is stopping his brain from working normally. |
| The alcohol has weakened his muscles. |
| The alcohol has damaged his heart. |

How well am I doing?

On track

I know some of the things I must do to stay healthy.

Aiming higher

I know some of the poisons I must avoid.

5 What does a life cycle explain?

- If living things did not reproduce, they would die out.
- Humans and plants have life cycles that are similar in some ways.

Every plant and animal (including humans) grows old and dies. They all reproduce while they are alive to keep the species going. Birds lay eggs which hatch into new chicks, and humans have babies. Plants make seeds that can grow into new plants.

How many babies do living things need to make to keep the species alive?

Humans have very few babies. Most of them will survive, be healthy and grow up.

If families only have two or three children it is enough to keep the human race going. If they have more than that, the population will grow.

This chestnut tree makes hundreds of seeds called conkers every year. Most die. Only a few make a new tree.

Even so, this makes enough new trees every year as old ones die out.

What are the life cycles of humans and plants?

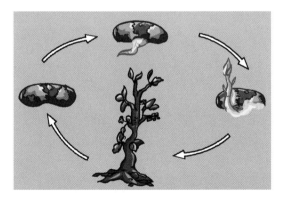

Plants and humans (and animals) grow up into adults. They replace themselves before they die.

Reproduction
Reproduction means making more copies of something – having babies, for example.

Cycle
Anything which has different stages happening over and over again is a cycle.

Test your learning

On track

1   Every plant and animal can reproduce. Humans do it by having babies, and flowering plants make seeds which grow into new plants.

human baby   acorns

(a)   What do birds make when they reproduce?

(b)   Explain why every living thing needs to reproduce.

(c)   Explain why a sea fish such as a cod has thousands of babies each year.

Aiming higher

2   Here are some stages in the life cycle of a flowering plant. They are in the wrong order.

| adult | fertilisation | germination | pollination | dispersal |

(a)   Write the names of the stages in the right order, starting with adult.

Here are four things that all living things can do.

| feed | grow | reproduce | move |

(b)   Which one is the main job of a flower?

How well am I doing?

On track

I can say why every living thing needs to reproduce.

Aiming higher

I can name the main stages of the life cycles of humans and plants.

6 How do flowers make seeds?

- Flowers have male and female parts.
- A seed is formed when pollen (male) fertilises the ovum (female).

The flower makes a plant's seeds. It has male parts called stamens, and the female part is called the carpel. Pollen from the stamen fertilises the ovum inside the carpel. This produces a seed, which is actually the beginning of a new plant.

Where are the male and female parts of a flower?

The female part is called the carpel. It is in the middle and contains the ovum. This is the plant's egg.

Surrounding the carpel are the male parts called stamens. The pollen is in the bags (called anthers) on the end of the stamens.

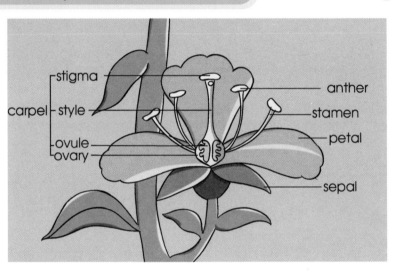

How do these help a seed start to form?

The first thing that has to happen is to move male pollen grains from the stamen to the stigma. This is called **pollination**.

Attracted by the colour and scent of the flowers, bees collect sugary nectar. Pollen sticks to the bee's body. It brushes off onto the carpel of the next flower the bee visits.

The ovum will turn into a seed after the pollen grain **fertilises** it.

- Once the pollen grain lands on the tip of the carpel, a little tube grows out of it and down to the ovum.
- Part of the pollen grain goes down the tube and joins up to the ovum.
- The ovum has been fertilised. Now it can turn into a seed.

Pollination

Moving pollen from one flower to the next.

Fertilisation

When part of the pollen grain joins the ovum.

Test your learning

On track

1   Here is a picture of a flower, which has been labelled with numbers. You can also
    see the names of the parts in a box next to the diagram.

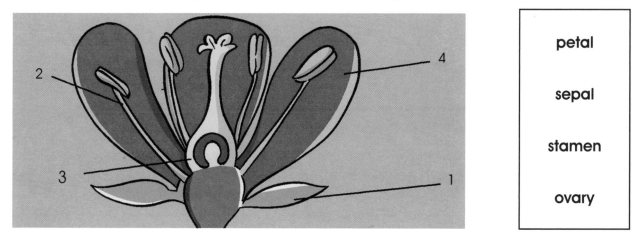

petal

sepal

stamen

ovary

(a) Write out the names of the flower parts and match up the right number to each one.

(b) Where is pollen made?

(c) How is pollen moved from one plant to another?

Aiming higher

2   Jaldev looked at a rose.

(a) Explain why this flower is brightly
     coloured and has a strong scent.

(b) What is the nectar for in a flower?

How well am I doing?

On track

I can name the parts of a flower.

Aiming higher

I can say how each part helps the plant
to reproduce.

7 How are seeds dispersed?

- Seeds need a good place to grow.
- Seeds are dispersed in three main ways.

If a seed falls straight down from its parent plant, it might find it hard to grow. The big parent takes all of the light and water. To get a better chance of living, the seed needs to land further away from its parent. Plants have three main ways of dispersing seeds.

Why do seeds need to be dispersed?

Seeds need warmth and water before they can start to grow. Once they get started they also need light. Only the seeds that get to a place where there are plenty of those things can survive.

If they just fall underneath their parent they will be shaded from the light, and their big parent might be taking all the water. They need to move further away!

What are the main ways seeds are dispersed?

Seeds are often inside fruits which help them to be dispersed.

By animals	**By the wind**	**By bursting open**

Many fruits taste nice. This horse is eating apples. The seeds come out in its droppings.	These seeds are carried by parachutes. They blow away in the wind to a new place.	These fruits dry out and burst open. When they explode the seeds are dispersed.

Dispersal
Scattering seeds away from the parents so they can grow better.

Fruit
The seed container.

Test your learning

On track

1   Here are four fruits. In the box are four ways that fruits can be dispersed.

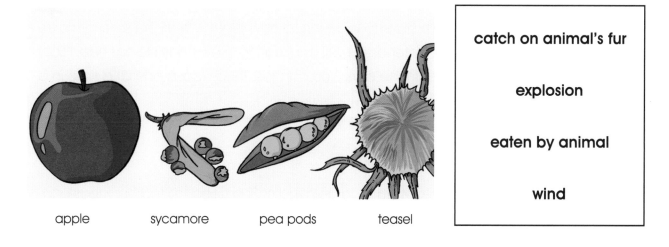

apple	sycamore	pea pods	teasel

> **catch on animal's fur**
>
> **explosion**
>
> **eaten by animal**
>
> **wind**

(a) Write the name of each fruit and its method of dispersal.

(b) Explain why seeds need to be dispersed from the parent.

Aiming higher

2   Here is a picture of a tomato. The 'pips' inside it are the seeds. Its seeds are dispersed mainly by animals.

(a) Why do tomatoes produce seeds?

(b) What clues tell you that tomato seeds are dispersed by animals?

(c) Explain how an animal would disperse the tomato seeds.

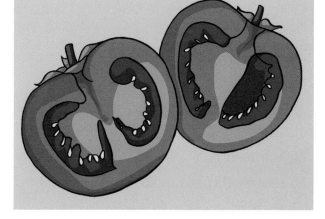

How well am I doing?

On track	Aiming higher
I can say why seeds need to be dispersed.	I can describe the different ways seeds can be dispersed.

8 What makes seeds germinate?

- Fair tests help you find out the best conditions for germination.
- Several tests are needed to make the results more reliable.

Germination is when a seed first starts to grow into a new plant. When the soil warms up in the spring the seeds soak up water and swell and start to grow. They do not need light – seeds germinate perfectly well in the dark. You can test these ideas one at a time in a **fair test**.

How can you set up a fair test?

Cecilia wants to see if plants need warmth to germinate. She needs to test her ideas one at a time to be fair. She tried two ways before she got it right. This is what she wrote.

The wrong way	The right way
I put one box of <u>dry</u> seeds in a <u>cold</u> fridge, and another with <u>moist</u> seeds on a <u>warm</u> radiator.	Then I put one box of <u>moist</u> seeds in the <u>cold</u> fridge, and another on a <u>warm</u> radiator.
Only the wet, warm ones germinated. Was it being cold or being dry that made some fail? I couldn't tell!	They were both moist, but only the warm ones germinated. Now I can see they must have warmth to germinate!

Why test many seeds?

Experiment 1	Number growing
1 seed in a cold place	0
1 seed in a warm place	0

Experiment 2	Number growing
10 seeds in a cold place	0
10 seeds in a warm place	8

Testing many seeds makes your results more reliable.

If you only test one seed, you could make a wrong conclusion. If your warm seed has failed, you could think warmth is not what they need.

If you test ten seeds, you can see the difference. Even though two warm seeds have failed, you still know that eight could germinate in the warmth.

Fair test

Testing your ideas one at a time.

Conclusion

What your experiment has found out.

Test your learning

On track

1 In this experiment Cheetah class were growing some seeds on damp cotton wool. They placed 50 seeds in each dish. There were three dishes.

- Dish 1 was kept in a fridge at 4°C.

- Dish 2 was kept in a dark outdoor shed at 18°C.

- Dish 3 was kept in a dark cupboard with a radiator at 25°C.

They counted how many of the seeds had germinated after one week.

(a) What idea were the children testing?

(b) Why did they have to make sure that every dish was kept moist and dark?

Aiming higher

2 Here are their results.

	Temperature	Numbers growing
Dish 1	4°C	0
Dish 2	18°C	32
Dish 3	25°C	40

(a) What conclusion could they draw from their results test?

(b) Explain why it was necessary to use so many seeds.

How well am I doing?

On track

I can make a test fair by changing one factor at a time.

Aiming higher

I can make my results more reliable by using many samples.

9 How do humans grow up?

- Humans pass through different stages.
- Children depend on their parents for much longer than other animals.

You started life as a baby. Now you are in your childhood. When you are in your teenage years, you will be an adolescent. Finally you will be an adult. As you get older you become less dependent on your parents. Other animals grow up much faster than humans.

What stages do humans pass through?

Name of stage	Description	Time taken
Baby	Small and helpless	0–18 months
Child	Still needs a lot of help. Can't have babies	18 months –11 years
Adolescent ('teenager')	Learning adult skills	11–18 years
Adult	Able to have a baby and look after it	18 years +

A newborn human baby relies on its parents for almost everything. It needs milk from its mother to drink and its parents have to do everything for it. It takes up to 18 years for the baby to learn to do everything for itself.

How long do children depend on their parents?

Human children depend on their parents longer than any other animal. We grow slowly, and there is so much to learn.

You will lawfully be an adult when you are 18, but your parents might still be helping you pay for university when you are 21!

Other animals grow up much faster. Many animals are adult after about two or three years. A cow can have her first calf when she is only two years old!

Adolescence

The teenage years when a child is turning into an adult.

Dependent

A person or animal who needs a lot of help to survive.

Test your learning

On track

1   Here are the stages of a human life cycle. Some statements are given in the box.

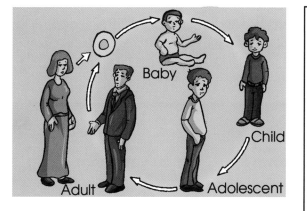

- Have to have everything done for them.
- Are old enough to look after their own babies.
- Are just turning into adults.
- Can walk and talk but need looking after by their parents.

(a) Write out the four stages, and match each one up to the right statement.

(b) Which stage is often called 'teenager'?

(c) A cat is an adult at three years old – at about what age do people become adult?

Aiming higher

2   Human beings are dependent on their parents for many years. Sally is 20 years old and her mum still gives her some money to help her at university.

(a) Why do humans stay dependent on their parents for such a long time?

(b) Name four things your parents do for you that help you survive and grow up.

How well am I doing?

On track

I can describe the main stages of human development.

Aiming higher

I can explain the long period of humans' parental care.

10 What is air?

- Air is a mixture of gases that you can feel when the wind blows.
- You can measure the mass of air accurately by averaging several measurements.

An empty bottle actually still has something in it – **air**! We forget all about air because we can't see it. It has no colour or smell and very little weight. It is a real substance though. We would die in a room without any air.

How do we know air is a real substance?

Air holds things up

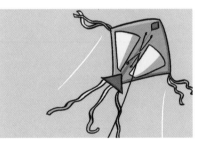

A kite is held up by the air.

Air slows you down

Air resistance slows a car down.

Air moves things along

The wind can make a sailing ship move very fast.

How do you measure the mass of air?

Air has a mass – but it is very small! You need very sensitive equipment to detect it.

When you weigh a full balloon, you will find that it has a very slightly bigger mass than an empty one.

You need to repeat your measurements. Air is so light that just a little breeze will make your readings different each time. You will need to take many readings and work out the average.

Test	Reading on balance
1	20.11 g
2	20.24 g
3	19.95 g
Average	**20.10 g**

Air resistance

The force of the air which slows down moving objects.

Repeated measurements

A reading taken several times to make the result more accurate.

Test your learning

On track

1   Annetta has chosen a bad day for a walk. It is so windy!

(a) What clues can you see in the picture that show that air is a real substance?

(b) What would happen to the aeroplane if there was no air?

(c) What does Annetta need the air for?

Aiming higher

2   Jolie and Dominic were trying to find out the mass of a blown-up balloon.

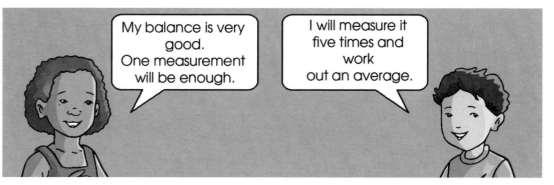

(a) Who do you think will have the most accurate answer?

(b) Explain how the balloon's mass might seem to be different in Dominic's five tests.

How well am I doing?

On track

I can give a few pieces of evidence to show that air is a real substance.

Aiming higher

I can explain why measurements need to be repeated a few times.

11 How much air is there in soil?

- Animals that live in the soil need air to breathe.
- You can measure the amount of air in the soil.

How do you think worms that live in the soil can breathe? There must be some air in soil. It is difficult to see how much air there might be, but you can measure it with an experiment. You need to do the experiment at least five times to get a good average.

Does soil have air spaces?

The ground feels quite hard when we walk on it. However, there are air spaces in between the soil particles. We know they are there because rainwater can soak away.

These air spaces are important for the animals in the soil. Animals in the soil need the oxygen in the air to breathe and stay alive.

How do you measure the amount of air in soil?

Cheetah class wanted to know if soil from different places contains different amounts of air.

They measured this by slowly pouring water into a beaker which was full to the top of dry soil. The water filled up the air spaces. They stopped when the water started to run over the top of the beaker. The volume of water they added was the same as the amount of air in the soil.

They tested each sample a few times and worked out the average.

How much water fits into a 200 cm³ beaker of soil?	
Type of soil	Volume of air
School field	40 cm³
Dominic's garden	60 cm³
Teacher's garden	50 cm³

Volume

The amount of space taken up by something.

1 cm³

A unit of volume. A litre contains 1000 cm³.

Test your learning

On track

1   Dominic wanted to know which had more air spaces – soil from the school field or soil from the woodland. He noticed the woodland soil seemed wetter, so he dried both types out completely before he started. Then he filled a 200 cm³ beaker to the top with some dry soil. He measured how much water he could add to the beaker before it ran out over the edge.

(a) Where does the water go when it is poured onto the soil?

(b) What do worms need the air spaces for?

(c) Why did Dominic have to dry out both samples completely before he started?

Aiming higher

2   Here are Dominic's results. He did the test five times.

Amount of water that can be added to 200 cm³ of dried-out soil		
	Garden soil	Woodland soil
Test 1	38 cm³	36 cm³
Test 2	40 cm³	32 cm³
Test 3	42 cm³	34 cm³
Test 4	41 cm³	33 cm³
Test 5	39 cm³	35 cm³

(a) Work out the average amount of air in each kind of soil.

(b) What conclusion could Dominic make from his results?

(c) Why did he measure each type of soil five times?

How well am I doing?

On track

I can say how animals can breathe under the ground.

Aiming higher

I can explain how to measure the amount of air in soil.

12 What gases do we use?

- Many gases are very useful to people.
- Their properties are well suited to the jobs they do.

People talk about 'gas' which they burn in their homes – that is a gas called **methane**. Doctors and dentists use 'gas' to put people to sleep – that could be one of a few different **anaesthetic** gases. Then there is the gas in fizzy drinks, and another in balloons.

What gases do we use?

- Party shops have cylinders of **helium** to fill party balloons.

- Fizzy drinks such as fizzy lemonade and beer have got **carbon dioxide** dissolved in them.

- Doctors use **anaesthetic gases** to put people to sleep for their operations.

- **Methane gas** is found under the North Sea and brought to our houses in pipes to help heat our homes.

How do you measure the amount of air in soil?

The patient is being given a mixture of anaesthetic gas, to put her to sleep, and **oxygen**, to keep her alive.

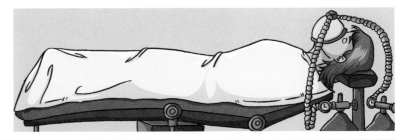

Name of gas	Useful properties
Helium	Lighter than air, so helium balloons float up to the ceiling.
Carbon dioxide	Dissolves in water and turns into 'fizz' when you open a bottle.
Methane	Burns really well.
Anaesthetic	Puts a patient to sleep so they feel no pain.

The properties of each gas make it suitable for the job it does for us.

Methane

The common name for gas we use in a cooker.

Anaesthetic

A substance used by doctors to put a patient to sleep. Used in operations.

Test your learning

On track

1   Here are the names of some common gases.

| anaesthetic gas | carbon dioxide | helium | methane | oxygen |

(a) Copy out this table and match up the name of each gas above to these statements.

used by doctors to help patients breathe

used by doctors to put people to sleep

makes bubbles in fizzy drinks

burned in gas boilers

used to make balloons float

Aiming higher

2   These are the properties of some gases.

Gas A burns really well.

Gas B is lighter than air.

Gas C makes patients unconscious.

Gas D forms bubbles in water.

Gas E is needed for a person to stay alive.

(a) Copy out this table and match up the letters A to E above with the correct gas.

anaesthetic gas

carbon dioxide

helium

methane

oxygen

How well am I doing?

On track

I can say what some common gases are used for.

Aiming higher

I can explain how the properties of the gases help them to do their jobs.

13 What is evaporation?

- All liquids evaporate to form gases.
- Some liquids evaporate faster at room temperature than others.

Any liquid left to stand will dry up after a while. This is because it has evaporated. When liquids evaporate they turn into a gas and often go into the air. Liquids with low boiling points evaporate faster than those with high boiling points.

What is evaporation?

Jolie noticed that all liquids dry out. For example, puddles of water seem to dry up. But the water has not really gone away; it has just changed into water vapour and gone into the air.

Perfume is the same. As it evaporates, the gas spreads out into the room. The air carries it to you.

How fast do liquids evaporate?

Washing hung up outside on the line could take a few hours to dry. It is much quicker in a tumble drier. The higher the temperature outside, the faster the water evaporates.

Not only water evaporates. Perfume evaporates faster than water because its boiling point is much lower.

Evaporate

When a liquid turns into a gas.

Water vapour

The proper name for the gas that liquid water turns into.

Test your learning

On track

1   Jaldev and Annetta measured a puddle in the playground. The rain had stopped and the puddle was drying up in the sunshine. They then measured the puddle every half an hour.

Time	Size of puddle
9:00	280 cm
10:00	140 cm
10:30	70 cm

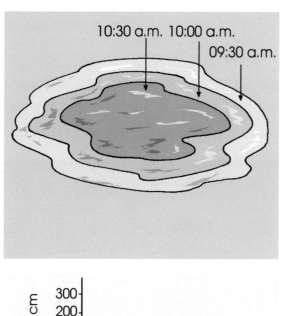

(a) Draw a graph to show how the puddle changed size.

(b) At what time do you think the puddle might have disappeared altogether?

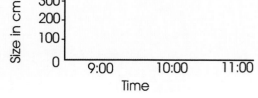

(c) What was happening to the water to explain why the puddle was drying up?

Aiming higher

2   Annetta looked up the boiling points of some household liquids on the Internet.

Liquid	Boiling point
water	100°C
paintbrush cleaner	160°C
olive oil	300°C
vinegar	118°C

(a) List the liquids in the table in the order of how fast they evaporate at room temperature. Start with the slowest first.

How well am I doing?

On track

I can explain what evaporation is.

Aiming higher

I can describe the link between boiling point and speed of evaporation.

14 How do solids, liquids and gases differ?

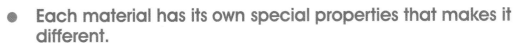

- Each material exists as a solid, liquid or gas.

- Each material has its own special properties that makes it different.

Each substance is either a **solid**, a **liquid** or a **gas** at room temperature. You can tell the difference between solids, liquids and gases by using some simple rules. The way you decide this is to look at their properties.

What are some examples of solids, liquids and gases?

Solids	Liquids	Gases
ice	water	steam
wood	oil	oxygen
iron	petrol	neon
rubber	custard	helium
stone	blood	methane

How fast do liquids evaporate?

Rule: Solids are hard and keep their own shape.

Rule: Liquids flow and take the shape of their container.

Rule: Gases can be squashed and they spread out to fill a whole room or container.

Hard
Doesn't change shape when you hit it.

Squash
To crush something into a smaller space.

RISING ★ STARS

Science Study Guide: Year 5

Answer Booklet

Unit		On track		Aiming higher
1 What is healthy eating?	**1 (a)** **(b)** **(c)**	Chicken leg Pasta Apple	**2 (a)** **(b)**	They contain vitamin C. To see if they had any effect.
2 What does your heart do?	**1 (a)** **(b)** **(c)**	Muscle Pumps blood around the body. Circulates	**2 (a)**	**Blood vessels** carry blood around. **Heart** pumps blood. **Blood** carries energy. **Rib cage** protects the heart.
3 Are exercise and pulse rate linked?	**1 (a)** **(b)** **(c)**	95 beats per minute 5 minutes So that the blood can take more oxygen and energy to the muscles.	**2 (a)** **(b)**	128 To take an average.
4 How do you keep healthy?	**1 (a)** **(b)** **(c)**	**Take more exercise** and your heart will grow stronger. **Stop smoking** and you will have less chance of lung cancer. **Eat less fatty food** and you will become less fat. **Don't drink alcohol** and you will have less chance of a car crash. Puts strain on your heart: blocks your blood vessels/stops the blood flowing. Smoking gives you heart disease and/or lung disease such as cancer/bronchitis.	**2 (a)**	The alcohol is stopping his brain from working normally.
5 What does a life cycle explain?	**1 (a)** **(b)** **(c)**	Eggs To keep the species going. Most of the babies die and only a few grow into adults.	**2 (a)** **(b)**	adult – pollination – fertilisation – dispersal – germination – adult Reproduce
6 How do flowers make seeds?	**1 (a)** **(b)** **(c)**	1 sepal 2 stamen 3 ovary 4 petal In the anthers on the end of the stamens. By pollination, often by bees and other flying insects.	**2 (a)** **(b)**	To attract insects. Nectar attracts pollinating insects.
7 How are seeds dispersed?	**1 (a)** **(b)**	Apple – eaten by animal Sycamore – wind Pea pods – explosion Teasel – catch on animal's fur To have a better chance of growing.	**2 (a)** **(b)** **(c)**	To keep the species growing. The tomato is nice to eat. The animal would eat the tomato – the seeds would not be digested – they would come out in the animal's droppings.
8 What makes seeds germinate?	**1 (a)** **(b)**	Do plants need warmth to germinate? To make the test fair – to keep everything but the temperature the same.	**2 (a)** **(b)**	Plants do need warmth to geminate. To make sure they come to a good (valid) conclusion.

Unit		On track		Aiming higher
9 How do humans grow up?	**1 (a)**	Baby – Have to have everything done for them. Child – Can walk and talk but need looking after by their parents. Adolescent – Are just turning into adults. Adult – Are old enough to look after their own babies.	**2 (a)** **(b)**	Humans grow slowly and there is much to learn. Suitable answers include: money/food/a home/ clothes/love/all the things I need
	(b) **(c)**	Adolescent 18		
10 What is air?	**1 (a)**	Air slows the parachute down. Air makes the clothes move. The umbrella feels a force. The aeroplane would not fly.	**2 (a)** **(b)**	Dominic It is hard to measure such a small weight and this means there will be errors in weighing.
	(b) **(c)**	To breathe in the oxygen it contains.		
11 How much air is there in soil?	**1 (a)** **(b)** **(c)**	Into the spaces in the soil. To breathe in the oxygen. To make his test fair.	**2 (a)** **(b)** **(c)**	Garden soil = 40 cm³ Woodland soil = 34 cm³ Woodland soil can soak up less water/contains less air than garden soil. To work out an average and make his results more reliable.
12 What gases do we use?	**1 (a)**	**Oxygen** is used by doctors to help patients breathe. **Anaesthetic** gas is used by doctors to put people to sleep. **Carbon dioxide** makes bubbles in fizzy drinks. **Methane** is burned in gas boilers. **Helium** is used to make balloons float.	**2 (a)**	Anaesthetic gas (Gas C) makes patients unconscious. Carbon dioxide (Gas D) forms bubbles in water. Helium (Gas B) is lighter than air. Methane (Gas A) burns really well. Oxygen (Gas E) is needed for a person to stay alive.
13 What is evaporation?	**1 (a)** **(b)** **(c)**	Suitable graph 11:00 a.m. It is evaporating into the air.	**2 (a)**	olive oil – paintbrush cleaner – vinegar – water
14 How do solids, liquids and gases differ?	**1 (a)**	Solids: iron/sponge/salt Liquids: water/milk/petrol Gas: air/carbon dioxide/helium	**2 (a)** **(b)** **(c)**	Rule 1 Rule 3 Rules 2 and 4
15 How do you change the state of water?	**1 (a)** **(b)** **(c)**	Liquid Gas Solid	**2 (a)** **(b)**	Melting Reversible
16 What makes water evaporate faster?	**1 (a)** **(b)** **(c)** **(d)** **(e)**	The warmer the room the faster the water evaporates. She would need a beaker, something to measure water with, a thermometer, water and rooms of different temperatures. The temperature of the room. The amount of water she starts with/the temperature of the water at the start. The time it takes for all the water to evaporate.	**2 (a)** **(b)**	The warmer the room the faster the water evaporates. Repeat the tests in each room several times – then take an average of each set of results.

Unit		On track		Aiming higher
17 How hot is boiling water?	**1 (a)** **(b)**	26°C 68°C	**2 (a)** **(b)** **(c)**	20°C About 103°C Graph showing the lemonade cooling.
18 What affects how fast ice melts?	**1 (a)** **(b)**	30°C The hotter room makes the ice melt faster and the water heats up to room temperature faster.	**2 (a)** **(b)**	Classroom 3 I think a warmer classroom will melt ice faster because it can give more heat to melt the ice.
19 What happens in the water cycle?	**1 (a)** **(b)** **(c)**	The rivers flowing into it. Rain Rivers and the sea.	**2 (a)** **(b)**	The Sun makes water vapour rise from the sea into the air. Water vapour forms as the clouds cool. Water vapour cools and forms water droplets. Water droplets fall to the ground as rain. Streams and rivers take water back into the sea. When the Sun makes water vapour rise from the sea.
20 What shape are the Earth, Sun and Moon?	**1 (a)** **(b)** **2 (a)**	Spherical Photographs As the Earth curves away from you the ships' masts gradually get lower and lower and finally disappear.	**3 (a)** **(b)**	Sun largest, then the Earth, then the Moon. The Moon is smaller than the Sun but it is much nearer to us.
21 How do shadows change?	**1 (a)** **(b)** **(c)**	The Earth spins once a day. The Sun appears to rise in the East. The Earth is **not** stationary. The Sun does **not** move across the sky. The Sun appears to move across the sky because of the spinning of the Earth.	**2 (a)**	8 a.m. = B 11 a.m. = C 3 p.m. = A
22 What causes day and night?	**1 (a)** **(b)** **(c)**	At A it is midday/12 noon. At B it is sunset. It will be midnight again in 24 hours.	**2 (a)** **(b)**	June 21st: sunrise 04:40 and sunset 21:38. December 22nd: sunrise 08:23 and sunset 15:51. March and September at the time of the equinoxes.
23 Why does the Moon appear to change?	**1 (a)** **(b)**	28 days At point A the moon is full, and at point B it will be a half moon.	**2 (a)** **(b)** **(c)** **(d)**	25th October, 23rd November, 22nd December. Quarter Moons in December = 1st, 15th and 29th – any two for a mark. 19th January in the next year. Two

Unit		On track		Aiming higher
24 What is a year?	**1 (a)**	It takes 365¼ days for the Earth to orbit the Sun. The Earth spins on its axis once a day. It takes approximately 28 days for the Moon to orbit the Earth. It takes 24 hours for the Earth to spin once on its axis.	**3 (a)**	The first picture is correct. The second one is wrong because the Sun is not at the centre of the Earth's orbit.
	2 (a)	A leap year has one extra day to take account of the fact that the Earth takes ¼ day longer to orbit the Sun each year. So every fourth year we add 4 x ¼ or one extra day.		
	(b)	2020, 2032		
25 How are sounds made?	**1 (a)** **(b)**	His larynx. Jaldev's voice makes the air and then the can vibrate – the vibrations travel along the string to the other can, then through the air to Annetta.	**2 (a)** **(b)**	Blow air into the trumpet, hit the drum, bow the violin string. Air column in the trumpet, violin string, drum skin.
	(c)	Sound travels best through a solid.		
26 How well does sound travel?	**1 (a)** **(b)** **(c)**	Air 8 x 250 m/2000 m Quieter	**2 (a)** **(b)**	The further away a sound is, the quieter it is. Two oscilloscope graphs: the loud sound at 1 m has a tall vibration; the soft sound at 100 m has a shorter vibration.
27 Which materials muffle sound?	**1 (a)** **(b)** **(c)** **(d)**	Apparatus used: different materials (newspaper, fur, cotton wool), buzzer which works by itself, box, sound meter. The box, the amount of material, the distance the sound meter is away from the box, the buzzer. The material inside the box. To see how loud the buzzer is without any material, so they could see the effect each material has on the loudness of the sound.	**2 (a)** **(b)** **(c)**	Fur, screwed-up newspaper, cotton wool. The results were not as the pupils predicted. dB or decibels To repeat their measurements several times and take an average of their results.
28 Can an instrument's pitch change?	**1 (a)** **(b)**	Low-pitched note. Draw a wave with the same height but many more vibrations (peaks and troughs) per second.	**2 (a)** **(b)**	D, A, B, C The pitch would get lower.
29 How do wind instruments work?	**1 (a)** **(b)**	The column of air in the tube. Trumpet and clarinet.	**2 (a)** **(b)** **(c)** **(d)**	Air The one with the most water/least amount of air in it (C). The water in the jar. The one with the smallest amount of water/greatest amount of air in it (D).

www.risingstars-uk.com

Test your learning

On track

1 Here are some substances.

| water | air | carbon dioxide | iron | sponge | salt | helium | milk | petrol |

(a) Copy out this chart and put each substance into the right column.

Solid	Liquid	Gas

Aiming higher

2 Here are some rules about solids, liquids and gases.

Rule 1: It keeps its own shape.

Rule 2: It spreads out to fill up any container it is in.

Rule 3: It cannot be squashed.

Rule 4: It can be squashed.

Rule 5: It can be poured from one container to another.

(a) Which rule is only true for solids?

(b) Which rule is true for liquids and solids?

(c) Which two rules are true for gases?

How well am I doing?

On track

I can recognise substances as solid, liquid or gas.

Aiming higher

I can understand and use the rules that tell solids, liquids and gases apart.

15 How do you change the state of water?

- Ice, water and water vapour are the three forms of water.

- Changing the temperature changes the state of a substance.

The **water** you drink is a liquid. Ice cubes are also water but in a solid form. Water vapour is the third state of water. When water changes from one form to another we say it is changing its state. Warming it up or cooling it down makes the state change.

What are the three forms of water?

Icebergs contain solid water

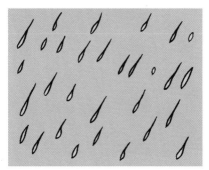

Rain is water in its liquid form.

Clouds contain water vapour – a gas.

How can you change the state of water?

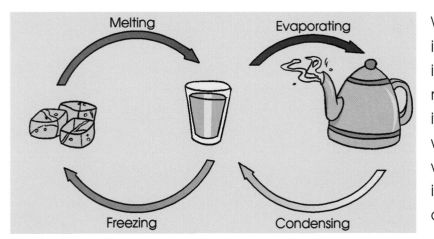

Warming solid ice makes it melt in water. Adding more heat makes it evaporate into water vapour. The whole process is reversible by cooling it down.

Look how the words 'melting', 'evaporating', 'condensing' and 'freezing' describe the changes of state.

States of matter

Describes if a material is a solid, liquid or gas.

Reversible

Capable of being changed back into its original state.

Test your learning

On track

1   The kettle is boiling. Dominic
    can see the steam coming out of
    the spout.

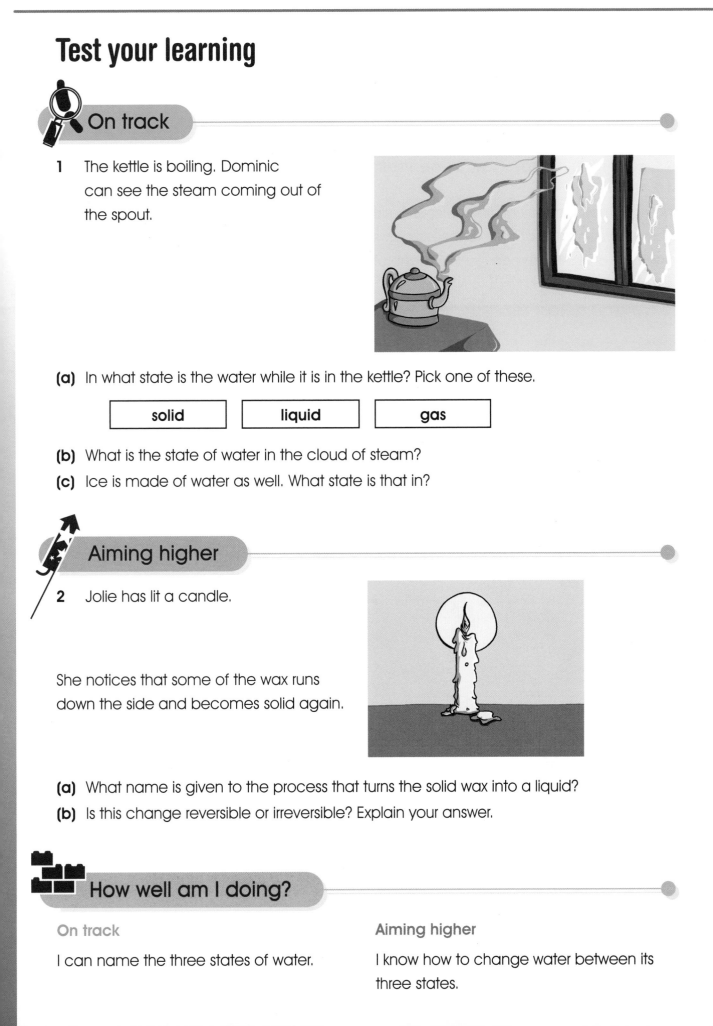

(a) In what state is the water while it is in the kettle? Pick one of these.

| solid | liquid | gas |

(b) What is the state of water in the cloud of steam?

(c) Ice is made of water as well. What state is that in?

Aiming higher

2   Jolie has lit a candle.

She notices that some of the wax runs
down the side and becomes solid again.

(a) What name is given to the process that turns the solid wax into a liquid?

(b) Is this change reversible or irreversible? Explain your answer.

How well am I doing?

On track

I can name the three states of water.

Aiming higher

I know how to change water between its
three states.

16 What makes water evaporate faster?

- When we plan an experiment, we make a prediction.
- Your results show how good your predictions are.

Water can evaporate at different rates. For example, puddles disappear more quickly when the weather is warm. If a breeze blows across a puddle it will dry up faster than when the air is still. You can test these and other ideas to see if they are true.

How do we test a prediction?

Dominic had four glasses containing water. Each glass was the same size but he then covered the top of each with a piece of card with different-sized holes. To make his test fair he had the same amount of water to start with for each glass and it was always at the same temperature.

Dominic predicted: "Water will evaporate faster in containers with larger openings."

He then measured the volume of water in cm^3 that had evaporated after four hours.

How can you change the state of water?

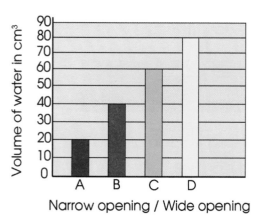

Look carefully at Dominic's results on this bar chart.

In the glass with the narrowest opening, only $20\,cm^3$ of water had evaporated. In the glass with the widest opening, $80\,cm^3$ of water had evaporated.

Dominic's prediction was right. The wider the opening the faster the water evaporates. This is because the surface area open to the air is greater.

Prediction

What you think is going to happen in an experiment.

Pattern

A set of results that fits into a scientific rule.

Test your learning

On track

1   Jolie wanted to test whether the warmth of the room made a difference to how fast water evaporated.

(a) Which of these predictions is most likely to turn out right in the end?

> "The colder the room the faster the water evaporates."

> "The warmer the room the faster the water evaporates."

> "The temperature makes no difference to how fast water evaporates."

(b) What would she need to do her best?

(c) What should she change as she carries out her test?

(d) What should she keep the same to make her test fair?

(e) What should she measure to collect her results?

Aiming higher

2   Jolie did her test and put her results in a table.

Classroom	Temperature of the room	Volume of water that evaporated after six hours
3	22°C	20 cm³
4	25°C	25 cm³
5	35°C	45 cm³
6	30°C	30 cm³

(a) Use the pattern in her results to say which of Jolie's predictions turned out to be true.

(b) What would you ask her to do to make her results more reliable?

How well am I doing?

On track

I can make a prediction about how an experiment will turn out.

Aiming higher

I can say whether the evidence supports a prediction.

17 How hot is boiling water?

- Graphs help you make a prediction.
- You need to do the test to see if your prediction is right.

When water is heated it warms up. You can draw a graph to show the water warming up. It starts at room temperature and rises steadily until the temperature reaches the boiling point of water (100°C). The temperature then stays the same.

Can you find a pattern in some data?

Time (minutes)	5	10	15	20	25	30	35	40
Temperature (°C)	20	25	30	35	40	45	50	55

Mr Rossi asked his class to look at this table. It shows the temperature of water which is being heated. He asked them to predict the next three results. They spotted the temperature was rising by 5°C every five minutes. Most predicted correctly that the next three readings would be 60°C, 65°C and 70°C.

Would their predictions still be true if they kept on heating the water even longer?

Does the pattern carry on forever?

Time (minutes)	65	70	75	80	85	90	95
Temperature (°C)	80	85	90	95	100	100	100

Mr Rossi asked if they thought this pattern would carry on. They said it would. They then looked at this completed table.

The pattern did not stay the same. Their prediction was not right. The temperature rose at a steady rate until it reached 100°C. Then it stayed at 100°C as long as it was being heated.

Mr Rossi said this was the boiling point of water.

Boiling point

The temperature at which a liquid boils.

Degrees Celsius

Units for measuring temperature invented by Anders Celsius.

Test your learning

 **On track**

1   Annetta heated some water and took the temperature at regular intervals. The water
    never boiled.

Time	2 minutes	4 minutes	6 minutes	8 minutes
Temperature	20	32	44	56

(a) What would the temperature be after three minutes?

(b) What would the temperature be after ten minutes?

Aiming higher

2   Annetta plotted
    a graph of some
    lemonade being
    heated. Lemonade
    is water containing
    lots of sugar.

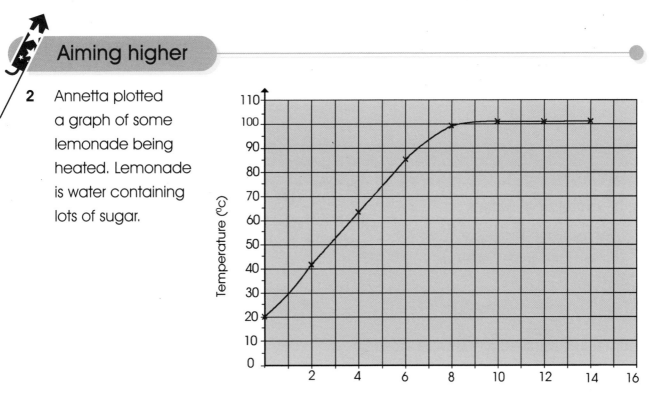

(a) What was the room temperature?

(b) What is the boiling point of lemonade?

(c) Draw a graph of what you think will happen when Annetta takes the heat away from
     the boiling lemonade.

How well am I doing?

On track

I can predict the next result on a graph by
looking at the first ones.

Aiming higher

I can change my prediction if the results
are not what I expected.

18 What affects how fast ice melts?

- Warming up solid ice melts it into liquid water.

- The temperature of the room affects how fast ice melts.

When ice warms up in a room it changes into liquid water. This happens at 0°C. The temperature of ice stays at 0°C until it all melts. Then the cold water warms up to the temperature of the room. The warmer the room, the faster the ice will melt.

What did Jaldev see when he let some ice warm up?

Jaldev put some ice cubes in water. He left this in his classroom and took the temperature of the mixture every five minutes. He then drew a graph of his results.

A The temperature is below 0°C. The ice slowly warms up.

B The temperature stays at 0°C (the melting point of ice) until all the ice melts.

C The water slowly warms up.

D Eventually the temperature of the water settles at room temperature, 22°C.

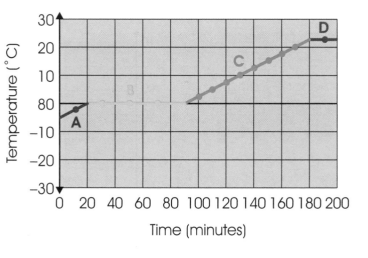

What affects how fast ice melts?

Ice melts because it takes heat from the room. What if the room was warmer to start with? Would a warm room make the ice melt faster than it would in a cold room? This table of results shows us the answer.

The warmer the room the faster the ice melted.

Temperature of the room	Time taken to melt an ice cube
18°C	10 minutes
20°C	8 minutes
22°C	6 minutes

Room temperature

The normal temperature of a room.

Table

A good way of organising results.

Test your learning

On track

1   Jaldev then did a second
    test. He put a similar ice/water
    mixture into a much warmer
    room.

Again he took the temperature of
the water every five minutes.

Here is the second graph he drew.

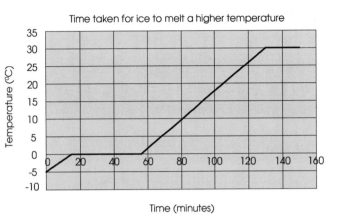

(a) What is the room temperature in this second room?

(b) Describe the difference the temperature in the second room has on the time taken
     for the ice/water mixture to warm up.

Aiming higher

2   Jaldev did a third test. He wanted to find the warmest classroom in the school.

He gave each class the
same-sized ice cube on a
plate. Each class measured
how long it took to melt.

Classroom	Time taken for the ice cube to melt
3	25 mins
4	30 mins
5	42 mins
6	28 mins

(a) Look at his results. Which classroom was the warmest?

(b) Describe how the temperature of the room affects the time taken for the ice cube to
     melt. Try using an "I think … because …" sentence.

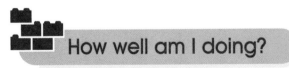

How well am I doing?

On track

I can describe what happens when ice
melts and warms up.

Aiming higher

I can explain why an ice cube melts faster
in a warm room.

19 What happens in the water cycle?

- The water cycle shows how water never stops moving from place to place.
- Changes of state help describe the water cycle.

All over the world rivers run into the sea. Have you ever wondered why the sea doesn't get deeper and deeper? What about the water in the rivers? Where does that come from? Will rivers ever dry up? The answer to all this is in the **water cycle**.

What happens in the water cycle?

The same water goes round and round in the water cycle. The cycle starts off in the clouds where water vapour turns into rain, filling up the rivers and lakes. The Sun dries up the lakes and the water finds its way back into the clouds again.

What changes of state happen in the water cycle?

Evaporation This is when the heat from the Sun makes liquid water (in the lake) change into water vapour (in the air).

Condensation This is when water vapour in the air gets so cold in the sky that it turns into liquid rain drops.

Freezing If it is really cold, the rain might freeze into hail or snow. The water in the lakes might freeze over.

Melting If the temperature is over 0°C, any ice that is around will melt into water.

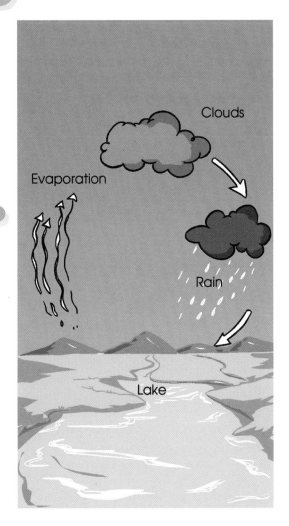

Water vapour

The name given to the gaseous form of water found in the air.

Precipitation

A more technical name to describe rainfall, hail or snow.

Test your learning

On track

1   This drawing shows part of the water cycle.

(a) The Sun keeps drying the sea up. What refills it?

(b) What keeps the streams from drying up?

(c) Where do the clouds get their water from?

Aiming higher

2   The statements below describe the water cycle. They are not yet in the right order.

> Water vapour forms as the clouds cool.
> Streams and rivers take water back into the sea.
> Water vapour cools and forms water droplets.
> **1** The Sun makes water vapour rise from the sea into the air.
> Water droplets fall to the ground as rain.

(a) Copy the five statements out in the right order to describe the water cycle. The first one has been marked with **1** to get you started.

(b) Where in the water cycle does evaporation take place?

How well am I doing?

On track

I can describe what happens in the water cycle.

Aiming higher

I can use changes of state to explain how the water cycle works.

20 What shape are the Earth, Sun and Moon?

- The Earth, Sun and Moon are spherical but are different sizes.
- From the Earth, the Sun and Moon look the same size.

Many centuries ago most people thought that the Earth was flat. Some still do so today! We now know it is spherical. What is the evidence for this? Only when we put cameras into space to photograph the Earth could we be certain that it was spherical.

How do we know the Earth, Moon and Sun are spherical?

Some people used to think the Earth was a sphere although they had no direct evidence. They said that ships disappeared over the horizon because the Earth curved. It kept on curving to eventually form a sphere.

In the last 50 years, scientists have gone into space. They have taken photographs of the Earth from all directions which show it is spherical. This is first-hand, convincing evidence that the Earth is a sphere.

Why do the Sun and Moon seem to be the same size?

Compared with the Earth the Sun is very large. Its diameter is about 100 times larger than the Earth. The Moon is much smaller. The diameter of the Moon is about a quarter of that of the Earth.

If the Sun were represented by a beach ball, the Earth would be a pea and the Moon a peppercorn.

In the sky, the Moon and the Sun look about the same size. Although the Sun is much larger it is also much further away, making it look about the same size as the Moon.

Sun
The only star in our solar system.

Earth
The planet we live on and the only one we know to contain life.

Test your learning

On track

1   On 12th April 1961, the Russian cosmonaut Yuri
    Gagarin became the first person in space.

(a) What shape is planet Earth?

(b) What evidence do you think Yuri Gagarin brought
     back to prove this?

2   Jaldev looks out to sea. Some big ships are disappearing over the horizon.

(a) Use the fact that ships disappear over the horizon to explain the Earth's shape.

Aiming higher

3   Jaldev looks at the sky. He sees the Sun and the Moon. They look about the same size.

(a) Put the Earth, Sun and Moon in order of their size, putting the larger first.

(b) Explain why the Sun and the Moon look the same size, even though they are not.

How well am I doing?

On track	Aiming higher
I can recognise the size and shape of the Earth, Sun and Moon.	I can explain why the Sun and the Moon look as if they are the same size.

21 How do shadows change?

- The Earth's spin makes the Sun appear to move across the sky.
- Shadows change in a regular way during the day.

Everybody says the Sun 'comes up in the morning'. They use the sunrise to describe what they see. It also seems to 'go down' in the evening. Is that what is really happening? It certainly looks like it but the truth is quite different.

Does the Sun really move across the Sky?

You might think the Sun moves across the sky, but this is not true. Think of it this way. Imagine you are driving in a car. You see houses move by but it is the car that is moving, not the houses. In a similar way the Earth moves, spinning on its axis once a day. This makes you think the Sun is moving across the sky. Actually, the Sun is still.

How do shadows change during the day?

The position and height of the Sun changes the length and position of shadows on the ground.

Mr Rossi's class did an investigation. They looked at how the Sun appeared to move across the sky and drew the shadows cast by a stick at different times. Here is what they discovered.

How does the Sun appear to move?
- Rises in the East and sets in the West.
- Is lowest at sunrise and sunset.
- Appears highest at midday.
- Moves in an arc across the sky.

What do the shadows look like?
- Are longest at sunrise and sunset.
- At 10 and 2 o'clock are the same length.
- Are shortest at midday when the Sun is highest.
- Always point away from the Sun.

Earth's axis
An imaginary line passing through the North and South poles of the Earth.

Midday
Usually regarded as 12 o'clock in the daytime.

Test your learning

On track

1   Mr Rossi gave his class some ideas to think about. Some are true, some are false.

> The Sun appears to rise in the East.
>
> The Sun moves across the sky.
>
> The Earth is stationary.
>
> The Earth spins once a day.

(a) Write down the two statements in the list that are true.

(b) Rewrite the wrong statements, making them true.

(c) Why does the Sun appear to move across the sky?

Aiming higher

2   These diagrams show the shadow sticks formed at different times of the day.

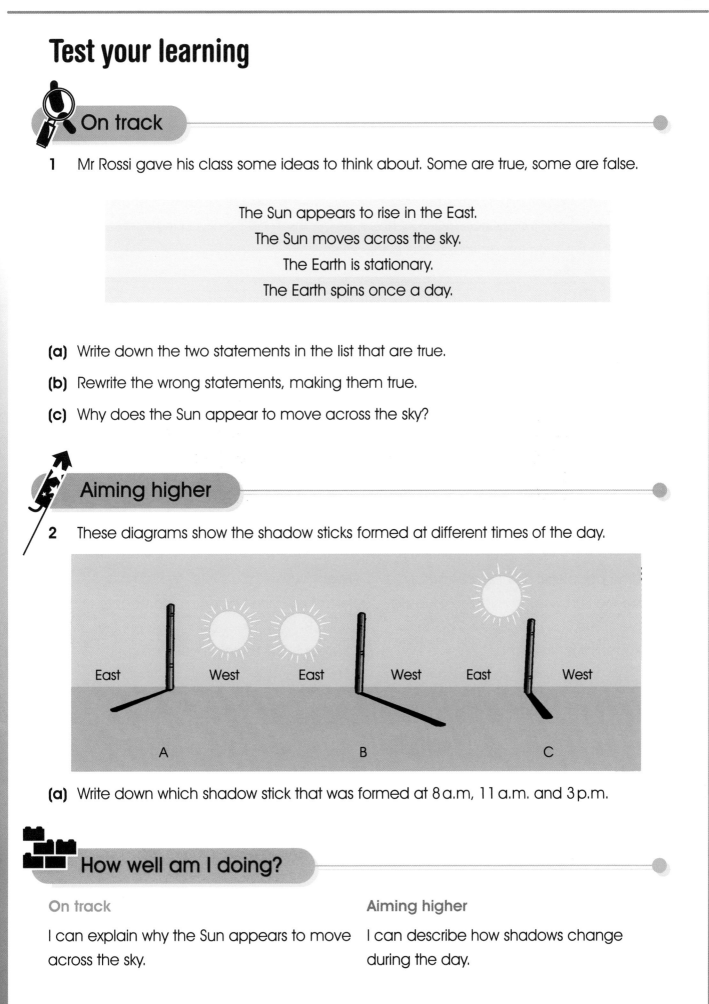

(a) Write down which shadow stick that was formed at 8 a.m, 11 a.m. and 3 p.m.

How well am I doing?

On track

I can explain why the Sun appears to move across the sky.

Aiming higher

I can describe how shadows change during the day.

22 What causes day and night?

- When it is daytime the point we are at on the Earth faces the Sun.

- There is a pattern in sunrise and sunset times during the year.

The Earth spins on its axis once every 24 hours. As it spins, most points on the Earth move through the light into the darkness, and back again.

How can you explain day and night?

Day and night are caused by the Earth spinning.

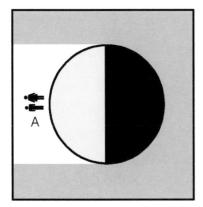

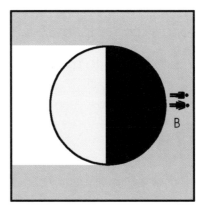

Day The people at point A are directly facing the Sun. It is the middle of the day for them. The Sun is straight up in the sky. This means it is noontime.

Night Twelve hours later the spinning (rotating) Earth has taken the people to point B. Now they cannot see the Sun. It is dark and it is midnight for them.

Do sunrise and sunset come at different times in the year?

- On midsummer's day (21st June) the Sun rises very early and sets very late. That is the longest day of the year.

- As summer ends and winter comes the days get shorter.

- On midwinter's day (21st December) the Sun rises very late and sets very early. That is the shortest day of the year.

- As winter ends and spring comes the days get longer.

Equinox

The dates in March and September when night and day are of approximately equal length.

Solstice

The dates in June and December when night and day are longest and shortest.

Test your learning

On track

1 This photograph shows the Earth. It is spinning anticlockwise.

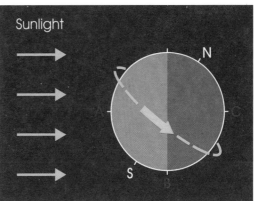

(a) What time of day is it at point A?

(b) If you were at point B would you see the sunrise, sunset or is it midday?

(c) At point C it is midnight. How many hours later would it be midnight again?

Aiming higher

2 Jaldev has drawn a graph of sunrise and sunset times for the city of Birmingham.

(a) Use the graph to estimate what time sunrise and sunset would be on the longest and shortest days of the year.

(b) In what two months of the year are day and night of equal length?

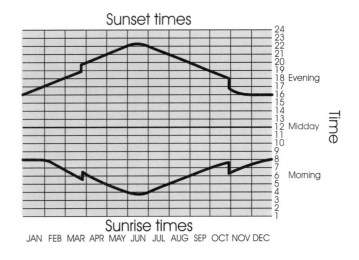

How well am I doing?

On track

I can say what causes day and night.

Aiming higher

I can understand the information in a sunrise and sunset graph.

23 Why does the Moon appear to change?

- The Moon orbits the Earth once a month.
- The shape of the Moon appears to change as it orbits.

Watch out for the Moon tonight. How much of it can you see lit up? Is it bigger or smaller than it was last night? Do you know when it will be full moon next? The changes that we see in the Moon are called its **phases** and they are caused by the Moon changing position in the sky.

How often does the Moon orbit the Earth?

We get our word *month* from the word Moon. It takes about one month for the Moon to go around the Earth once. It is in fact 29.5 days, so there are about 13 Moon cycles in a year, not 12.

How does the Moon look at different times of its monthly orbit?

A full Moon (Wed 9th): The Moon at the first point in its orbit. The Sun and Moon are opposite to each other. You see a bright disc.

A waning Moon (Wed 16th): The Moon has made a quarter turn. The left-hand quarter is lit brightly. Sunshine hits it sideways on.

A new Moon (Wed 23rd): The Moon is now halfway round its orbit. The Sun is more or less behind it so it looks dark.

A waxing half Moon (Wed 30th): When the Moon is three-quarters of the way around its orbit. The right-hand quarter is lit brightly. The sunshine is hitting it side-on.

Waxing

Growing.

Waning

Getting smaller.

Test your learning

On track

1 The picture shows the Moon making one complete orbit of the Earth. The Sun rises from the left.

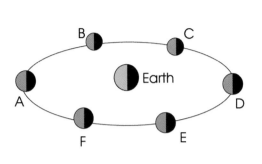

(a) How many days does it take for the Moon to make one complete orbit of the Earth?

(b) What shape of Moon would you see from the Earth when it is at point A and point D?

Aiming higher

2 The calendar shows the phases of the Moon over three months. On the 11th October there was a new Moon.

Look at the pattern.

		October						November						December				
Mon		2	9	16	23	30		6	13	20	27		4	11	18	25	
Tue		3	10	17	24	31		7	14	21	28		5	12	19	26	
Wed		4	11 ●	18	25 ○		1	8	15	22	29		6	13	20	27	
Thur		5	12	19	26		2	9 ●	16	23 ○	30		7	14	21	28	
Fri		6	13	20	27		3	10	17	24		1	8 ●	15	22 ○	29	
Sat		7	14	21	28		4	11	18	25		2	9	16	23	30	
Sun	1	8	15	22	29		5	12	19	26		3	10	17	24	31	

Key   ● = New Moon

○ = Full Moon

(a) On what dates was there a full Moon?

(b) Give two dates when there was a quarter Moon in December.

(c) What will be the date of the next full Moon?

(d) How many orbits of the Earth did the Moon make between 11th October and 8th December?

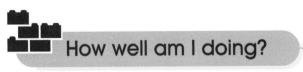

How well am I doing?

On track

I can say how often the Moon orbits the Earth.

Aiming higher

I can explain why the shape of the Moon changes during the month.

24 What is a year?

- The Earth moves around the Sun in a roughly circular orbit.
- It takes one year for the Earth to orbit the Sun once.

Why is a year about 365 days? The answer is to do with the way the Earth orbits the Sun. Your next birthday will come round when the Earth gets back to where it started on your last birthday. Even though the Earth travels a long way you can time it very accurately.

What does the Earth's orbit look like?

The Earth's orbit is nearly circular with the Sun at the centre.

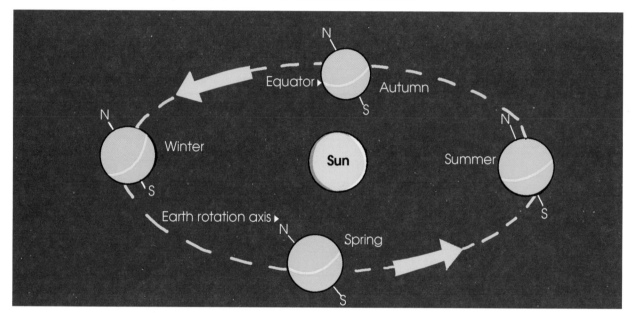

How long does one orbit take?

Although the Earth rushes through space at a speed of 67,000 miles per hour, it takes a whole year for it to orbit the Sun.

We normally think of a year from one birthday to the next as being 365 days. Actually it is a tiny bit more than that – it is $365\frac{1}{4}$ days. Every four years those extra $\frac{1}{4}$ days add up to a whole day, and we then have a year of 366 days. This is called a **leap year**.

Orbit

The circular path that one astronomical object makes around another.

Year

The period of time it takes the Earth to make a single revolution around the Sun.

Test your learning

On track

1 Jaldev has made a short quiz for his classmates.

1 **The number of days it takes for the Earth to orbit the Sun.**	1	7	24	28	$365\frac{1}{4}$
2 **How many times the Earth spins on its axis in one day.**	1	7	24	28	$365\frac{1}{4}$
3 **The number of days it takes for the Moon to orbit the Earth.**	1	7	24	28	$365\frac{1}{4}$
4 **The number of hours it takes for the Earth to spin once on its axis.**	1	7	24	28	$365\frac{1}{4}$

(a) Copy Jaldev's quiz. Circle the correct answers.

2 The years 2008, 2012 and 2016 are all leap years. They all have an extra day, 29th February.

(a) Explain why a leap year is longer than every other year.

(b) Which of these years will be leap years?   **2027   2020   2023   2032**

Aiming higher

3 Jaldev is unsure of how the Earth orbits the Sun. Only one of his diagrams is correct.

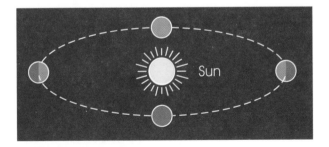 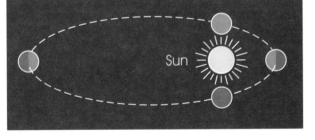

(a) Draw the orbit which is correct. Explain what is wrong with the other orbit.

How well am I doing?

On track

I can explain what the Earth's orbit looks like.

Aiming higher

I can say how long it takes for the Earth to orbit the Sun.

25 How are sounds made?

- You hear sounds when sound vibrations enter your ears.
- Different materials vibrate in different musical instruments.

Have you ever wondered how sounds are made? The idea is simple. If something vibrates, a sound is made. You hear the sound when the **vibrations** enter your ear. Musical instruments make a sound when strings, skins or columns of air vibrate.

How do you hear the sounds?

Sounds can travel through solids, liquids or gases. They travel best through solids and least well through gases such as air. You hear sounds when they enter your ear.

Jaldev hears sound underwater.   Sounds travel through air.   Sounds travel through
                                                                             solid and air.

What vibrates in musical instruments?

When objects **vibrate**, they make a sound. How can you show this?

Drums make a sound when the skin vibrates. If you put some rice grains on it they will jump up and down in time with the vibrations of the skin.

When guitar strings vibrate you hear a sound. If you put some chalk dust on one of the strings the dust will vibrate off.

Jaldev feels his larynx vibrating as he speaks.

Vibration	Sound
When an object or material moves back and forth around a fixed point.	A form of energy which can be detected by your ears or a sound meter.

Test your learning

 On track

1   Jaldev is talking to Annetta through the model telephone they have made.

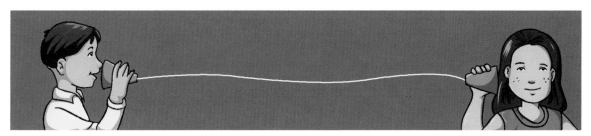

(a) What vibrates when Jaldev speaks?

(b) Explain how Annetta hears Jaldev speaking through the telephone.

(c) Which of these statements is true?

> Sound cannot travel through a solid.
>
> Sound travels faster through a gas than a liquid.
>
> Sound travels best through a solid.

Aiming higher

2   Here are some musical instruments. They each make a different sound.

(a) Explain how you would make each of these instruments produce a sound.

(b) What vibrates in each of them when they make a sound?

How well am I doing?

On track

I can explain how sounds are made.

Aiming higher

I can identify what vibrates in a range of musical instruments.

26 How well does sound travel?

- Sound travels quickly through air.
- The loudness of sounds depends on how far away they are.

Sound travels through air at about 250 m every second (770 mph). A simple experiment lets you find this out. A loud sound up close to you will sound softer the further away it gets until you can't hear it any more.

How can you tell how far away a sound is?

Annetta wanted to find out how fast sound travelled though air. She went into the playground and stood 125 m away from the school wall. She clapped her hands and heard an echo. Dominic timed this. He said there was one second between her clapping and them hearing the echo.

As sound travels 250 m (2 x 125 m) in one second, the speed of sound is 250 m per second.

What happens when sounds get further away?

Have you noticed that the sound of a car gets quieter as it gets further away from you?

A loud sound makes air vibrate violently. You cannot see this. However, a special device called an oscilloscope reveals the pattern of the sound.

You can use this to look at what happens to the loudness of a sound the further away it gets.

A loud sound close to you shows a big wave height. The same sound further away has a smaller wave height and sounds softer.

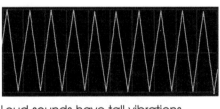

Loud sounds have tall vibrations.

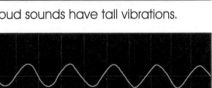

Softer sounds have shorter vibrations.

Oscilloscope
A scientific device which shows the shape of sound vibrations.

Amplitude
The height of a sound wave shown on an oscilloscope screen.

Test your learning

On track

1   Annetta saw the lightning. She heard the thunder eight seconds later.

    She knew she could see the lightning the instant it happened.

(a) What is the sound of the thunder travelling through?

(b) Use the idea that sound travels 250 m in a second to work out how far away the thunderstorm is.

(c) Would the thunder get louder or quieter as it moved away from you?

Aiming higher

2   Annetta wanted to know if she could hear Dominic's whistle at different distances. She stands different distances away from him while he blows his whistle hard. She writes down what she hears and looks at the sounds on her oscilloscope.

Distance between Dominic and Annetta	Sound Annetta hears
1m	extremely loud
25m	very loud
50m	loud
100m	quieter
200m	very quiet
300m	nothing

(a) Write a rule explaining how the loudness of the sound depends on the distance between Annetta and Dominic.

(b) Draw what Annetta might see on the oscilloscope when Dominic is 1 m and 100 m away.

How well am I doing?

On track

I can say how fast sound travels through air.

Aiming higher

I can say how the loudness of the same sound changes with distance.

27 What materials muffle sound?

- Fair tests help you see which materials muffle sound best.
- Test results help prove or disprove your predictions.

Have you noticed how the same sound differs from room to room? Soft materials in your house such as carpets or curtains **muffle** or absorb the sound. What is a loud sound in one room might not be in another. You can test to see which materials do this well.

How did Cheetah class test different materials?

Cheetah class wanted to test which materials muffled sound best.

They put a loud buzzer in a box. It was switched on from outside the box. On top of the box was a sound meter. This could measure the loudness of the buzzer inside the box.

First of all they measured the loudness of the buzzer in the box with nothing else in it. Then they put different materials in the box, one at a time, and measured the loudness of the buzzer again.

A sound meter measures the loudness of the sound in decibels (dB).

What predictions did the pupils make?

"The newspaper will not be very good because it is too thin."

"Fur will be very good because it is used in ear muffs."

"Cotton wool will be good because it is very soft."

The class thought fur would be the best and newspaper the worst at stopping sound getting out of the box. Their results would help them find out which prediction was best.

Muffle
To absorb or dull sound and prevent it entering your ears.

Insulator
A material that stops or reduces heat, light or sound passing through.

Test your learning

On track

1 Think about how Cheetah class planned and thought about their investigation.

(a) Make a list of all the apparatus they used.

(b) What factors do they keep the same in the test?

(c) What factor do they change in the test?

(d) Why did the class do the first test without any other materials in the box?

Aiming higher

2 Cheetah class made a table of their results.

Type of insulating material	Sound level (dB)
No insulating material	65
Cotton wool	44
Fur	28
Screwed-up newspaper	34

(a) Put the materials in the order of the best sound insulators. Put the best first and the worst last. Did the results turn out as the pupils predicted?

(b) What units are being used to measure the sound level?

(c) What would you ask the pupils to do to make their results more reliable?

How well am I doing?

On track

I can describe how to set up a fair test.

Aiming higher

I can say whether the evidence supports a prediction.

28 Can an instrument's pitch change?

- Every sound has a pitch which can be measured.
- The pitch of musical instruments can be raised or lowered.

The **pitch** of sounds is easily changed on musical instruments. Small drums and those with tighter skins have a higher pitch than larger ones. Guitar notes are higher when a string is shortened or tightened. Thick strings have a lower pitch than thin ones.

How do you measure the pitch of a musical instrument?

Oscilloscopes are special machines that show you what the sound vibrations look like. A microphone attached to the oscilloscope picks up sound. The vibrations are displayed on a screen. High-pitched sounds have more vibrations per second than low-pitched sounds.

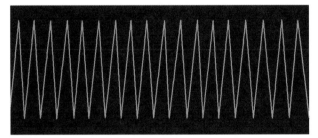
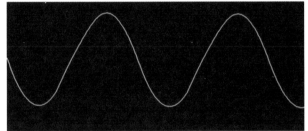

High-pitched notes have high frequencies or many vibrations every second.

Low-pitched notes have low frequencies or few vibrations every second.

How does Jaldev change the pitch of his instruments?

The tuners on Jaldev's guitar help him adjust the pitch of each note. As he tightens each string the pitch gets higher. As he loosens them the pitch gets lower. He adjusts the pitch of each string to the right level.

Jaldev also has a drum set. He has to tune it when he plays in his band. Screws on the side of the drum tighten or loosen the skin. If he needs the pitch to be higher, he tightens the skin.

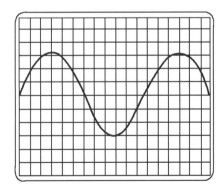

Pitch

Describes how high or low a sound is.

Frequency

The number of times a wave vibrates in a second.

Test your learning

On track

1   Jaldev plucks the thickest string on his guitar. The note is picked up by a microphone and shown on an oscilloscope screen. This is what it shows. He then plucks the thinnest string.

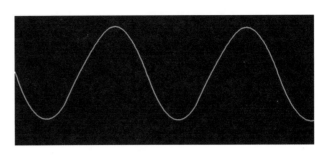

(a) Does the thickest string make a high- or a low-pitched note?

(b) Draw what the screen would look like when the thinnest string was plucked and the note was the same loudness.

Aiming higher

2   Here are Jaldev's drums. The skins are the same tightness but the pitch of their notes is different.

(a) Put the drums in order of the pitch of the notes they make. Put the highest-pitched drum first and the lowest-pitched last.

(b) What would happen to the pitch of each drum if you loosened each skin?

How well am I doing?

On track

I can explain how you measure the pitch of a sound.

Aiming higher

I can describe how to change the pitch of a musical instrument.

29 How do wind instruments work?

- Sound can be made by making air vibrate.
- The pitch of wind instruments can be raised or lowered.

A recorder is basically a long tube of air. It makes a sound when you blow vibrating air into it. You can change the pitch of the note by changing the length of the air inside the tube. The longer the tube of vibrating air the lower the pitch of the note.

What instruments work by blowing vibrating air into them?

You play brass instruments like the trumpet and trombone by blowing through your lips into the mouthpiece.

A clarinet has a reed in the mouthpiece that buzzes when you blow.

It is this blowing or buzzing that starts the air vibrating so the instrument makes its sound.

How do you change the pitch of a recorder note?

The length of the column of air inside a recorder is changed by closing up the holes with your finger. The lowest note on this recorder (bass C) sounds when all the holes are covered and there is a long air column. The highest note is made by the shortest air column, when no holes are covered.

The pitch gets lower as the air column gets longer.

Rule

A general pattern that can be used in different situations.

Note

A musical sound of a certain pitch.

Test your learning

On track

1   Annetta thought about how her recorder works.

(a)   What vibrates in a recorder to make a sound?

(b)   From this list, pick two other instruments that work in a similar way.

| trumpet | violin | piano | clarinet |

Aiming higher

2   Annetta and Jaldev had four bottles filled up to different levels, labeled A–D. They blew across the tops of the bottles and listened to the pitch of each note.

Then they tapped each bottle with a small stick. Again they listened to the pitch of each note.

Remember – not only air vibrates.

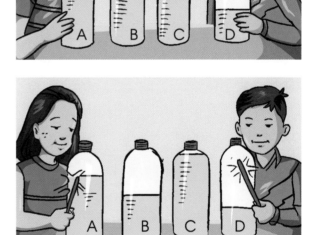

(a)   What was vibrating when Jaldev and Annetta blew across the bottle tops?

(b)   Which bottle would make the highest-pitched note when you blow across it?

(c)   What was vibrating when Jaldev and Annetta tapped each bottle with a stick?

(d)   Which bottle would make the highest-pitched note when it is tapped with a stick?

How well am I doing?

On track

I can explain how sounds can be made by air vibrations.

Aiming higher

I can describe how to change the pitch of a wind instrument.

Index